My Little Plastic Bag

Text and Concept Sam Love

Illustrations Samrae Duke

Published by Sam Love
Contact sam@samlove.net

Text Copyright © 2016 Sam Love
Illustrations Copyright © 2016 Samrae Duke
All rights reserved.

ISBN:1534622640
ISBN-13:9781534622647

The little plastic bag sparkles in the sunlight as Amy throws it from the car window

It floats down to the roadside where no one picks it up

On county mowing day big spinning blades cut the grass

chopping the little plastic bag into small shreds

The next rain carries the plastic pieces into the ditch

Then to the larger flooded stream

then to the tidal marsh

After the marsh it goes into the ocean

In the sea the sun and waves beat the plastic into tiny bits

Schools of small fish mistake the plastic for their food of fish eggs and plants

Large fish then eat the small fish which concentrates the plastic chemicals in the big fish and it can make them sick

Larger plastic pieces can also choke sea animals like turtles, fish and birds

A fisherman catches the big fish

He sells his catch to the fish market

Amy's family goes to the market and buys the big fish. The fish seller puts the fish in a new plastic bag.

They carry the fish home and her dad cooks it for dinner

Amy's family eats the big fish which has eaten the small fish that ate pieces of her plastic bag and this brings her
plastic bag back home

Amy may not realize she is eating the chemicals from her plastic bag, but now you do.

Now you know when you throw something away it can become part of a cycle in Nature.

Discussion Guide

How much plastic is in the ocean?

If we continue dumping plastic in the water, the world's oceans will soon have more plastic than fish. A report from the World Economic Forum and the Ellen MacArthur Foundation reports that there are over 165 million tons of plastics in the ocean today. The ocean is expected to contain 1.1 tons of plastic for every 3 tons of fish by 2025, and by 2050, more plastic than fish by weight. In other words, in a few decades, plastic trash in the ocean will outweigh all the fish in the sea.

Worldwide, nearly two million single-use plastic bags are used each minute. Plastic bags have consistently been reported as one of the most common forms of ocean litter. For every square mile in most of the ocean, there are about 46,000 pieces of plastic in it and some estimates say it can take plastic 1,000 years to degrade.

Is plastic getting into the ocean's food chain?

Tiny marine zooplankton are ingesting microplastic particles at an alarming rate, according to a study by Dr. Peter Ross, head of the Ocean Pollution Research Program at Vancouver Aquarium Marine Science Centre. Starting near the base of the aquatic food web, the plastics are radiating up the chains of predators and prey, finally accumulating in important food fish, such as salmon, that we eat.

Plastic acts like a sponge and soaks up other toxins from outside sources before entering the ocean. These chemicals are ingested by animals in the ocean and then we consume the contaminated fish.

About 44 percent of all seabirds eat plastic, apparently by mistake, sometimes with fatal effects. Marine species are known to swallow plastic bags, which resemble jellyfish in mid-ocean and they also face the invisible threat of toxic, plastic-derived chemicals.

Is all the plastic floating on top of the ocean?

An article in the journal Environmental Research by oceanographer and chemist Charles Moore, of the Algalita Marine Research Foundation gives the example of Styrofoam breaking down and the tiny polystyrene components start to sink because they're heavier than water.

"We knew ten years ago that plastic could be a million times more toxic than the seawater itself," because plastic items tend to accumulate a surface layer of chemicals from seawater, Moore said. "They're sponges."

Moore worries about the plastic-derived chemicals' potential damage to wildlife. The chemicals can potentially cause cancer in humans, he said, and simpler life-forms "may be more susceptible then we are."

Pollutants also become more concentrated as animals eat other contaminated animals—which could be bad news for us, the animals at the top of the food chain. (For more information read *National Geographic* magazine's "The Pollution Within.")

So the problem is not just the floating plastic we can see, but the tiny bits of plastic below the surface.

Is plastic in the ocean affecting humans?

In a study published in an issue of the journal *Science Advances*, plastics and other environmental pollutants found in fish are obstructing the human body's natural ability to expel harmful toxins.

There are different ways that plastic is dangerous for humans. Diethylhexyl phthalate (DEHP), contained in some plastics, is a toxic carcinogen. Other toxins in plastics are directly linked to cancers, birth defects, immune system problems, and childhood developmental issues.

Some plastics we know are toxic, such as #3, which is also known as PVC or poly vinyl chloride. PVC contains phthalates and heavy metals, and creates dioxins when it burns. Other plastics contain Bisphenol-A (BPA), which has been identified as a chemical that disrupts hormones. Plastics can contain thousands of possible additives, and manufacturers are not required to disclose their manufacturing recipes.

Rolf Halden, associate professor in the School of Sustainable Engineering at Arizona State University has studied plastic's adverse effects on humans. He has concluded it is almost impossible to determine the health effects on humans because plastic contamination is so globally spread that there are almost no unexposed subjects.

What can we do about the plastic getting into our water?

One of the most effective things we can all do is to be responsible for our trash. When possible, avoid buying products packaged in plastic. Always recycle plastic after using it. At the store, request a paper bag instead of plastic, or bring a reusable bag. Use a reusable water bottle, and of course don't litter.

Support efforts to regulate or tax the free distribution of plastic bags.

A recent study from the Earth Policy Institute, an independent non-profit environmental organization based in Washington D.C., reported there are 133 cities and counties in the United States that have passed a bag regulation ordinance. This includes 88 jurisdictions in California, the entire state of Hawaii and three counties on the Outer Banks of North Carolina. Two dozen other countries have a total ban or some other initiative to reduce plastic bag waste.

The Surfrider Foundation tracks reports on all jurisdictions that ban or tax single use plastic bags. Check out their web site at www.surfrider.org.

We can make a difference!

Activity

Make a Reduce, Reuse, and Recycle chart
Challenge the child to come up with ideas about what plastic can be reduced, reused and recycled.

The Original Poem

This book is based on Sam Love's poem
"The Downstream Loop" published
in his poetry collection "Converging Waters"
available on Amazon. It was later published
in Duke University's "Eno" magazine.

The Downstream Loop

Sparkling in the sunlight
the little plastic bag sails
from the mindless driver's hand
to drift among roadside weeds

No one bothers to retrieve it
and on county mowing day
whirring blades cut a grass swath
shredding the bag into gossamer slivers

The next thunderstruck downpour
carries the shreds through the watershed
to the larger boiling stream
to the tidal marsh
to the Atlantic ocean

The Gulf Stream's sun and waves
pulverize the slivers into tiny bits
creating a perfect culinary delicacy
for large schools of filter feeders
Small fish that mistake plastic globules
for aquatic eggs and plankton

Larger fish like Sea Trout and Tuna
cut a swath through the schools
and devour the tiny fish, concentrating
the petrochemicals up the food chain

For dinner we purchase the wild-caught Tuna,
let the fish monger filet the toxin-laden flesh,
pack it in ice, and store it in a virgin plastic bag
A bag that completes the ecological cycle

Contributors

Special thanks to the following people whose generous contributions through Indiegogo helped make this book possible:

Marcia Ostendorff
Nina Powell
Hunter and Eiko Brumfield
James Gaston
Kathy Long
Pamela Schiller
Ed and Jeri ODowd
Boyd Gatlin
Alexander and Robin Frelier
Richard Morgan

About Sam Love the author

Sam is a writer living in New Bern, NC. He was on the national staff of the first Earth Day and has worked over the years with numerous environmental campaigns. He worked professionally as a media producer. His articles on energy and the environment have been published in magazines such as Smithsonian and now he writes poetry that highlights the cracks in our culture. He has two published novels, Snap Factor, and Electric Honey. He recently published a poetry book Converging Waters.

He teaches yoga, and is one of the organizers of the First Tuesday Poetry Open Mike in New Bern. He views his Social Security check as a grant to the arts. His poetry has been published in numerous publications including Kakalak, Slippery Elm, The Lyricist and other publications. His environmental poems have been featured in the 2015 and 2016 issues of Eno Magazine published by Duke University.

About Samrae Duke the Illustrator

Samrae Duke is a professional illustrator whose work has appeared in numerous publications and juried art exhibitions. She has a BFA in Studio Art with a concentration in Illustration from East Carolina University School of Art and Design.

Lightning Source UK Ltd.
Milton Keynes UK
UKHW050241160719
346203UK00018B/697/P